ISBN 978-1-332-12534-0
PIBN 10288078

Forgotten Books is a registered trademark of FB &c Ltd.
Copyright © 2015 FB &c Ltd.
FB &c Ltd, Dalton House, 60 Windsor Avenue, London, SW19 2RR.
Company number 08720141. Registered in England and Wales.

For support please visit www.forgottenbooks.com

1 MONTH OF
FREE
READING

at

www.ForgottenBooks.com

By purchasing this book you are eligible for one month membership to ForgottenBooks.com, giving you unlimited access to our entire collection of over 700,000 titles via our web site and mobile apps.

To claim your free month visit:
www.forgottenbooks.com/free288078

English
Français
Deutsche
Italiano
Español
Português

www.forgottenbooks.com

Mythology Photography **Fiction**
Fishing Christianity **Art** Cooking
Essays Buddhism Freemasonry
Medicine **Biology** Music **Ancient**
Egypt Evolution Carpentry Physics
Dance Geology **Mathematics** Fitness
Shakespeare **Folklore** Yoga Marketing
Confidence Immortality Biographies
Poetry **Psychology** Witchcraft
Electronics Chemistry History **Law**
Accounting **Philosophy** Anthropology
Alchemy Drama Quantum Mechanics
Atheism Sexual Health **Ancient History**
Entrepreneurship Languages Sport
Paleontology Needlework Islam
Metaphysics Investment Archaeology
Parenting Statistics Criminology
Motivational

EUGENICS

THE SCIENCE OF HUMAN IMPROVEMENT
BY BETTER BREEDING

BY

C. B. DAVENPORT

CARNEGIE INSTITUTION OF WASHINGTON
DIRECTOR, DEPARTMENT OF EXPERIMENTAL EVOLUTION,
COLD SPRING HARBOR, N. Y.
SECRETARY, COMMITTEE ON EUGENICS, AMERICAN BREEDERS' ASSOCIATION

NEW YORK
HENRY HOLT AND COMPANY
1910

THE QUINN & BODEN CO. PRESS
RAHWAY, N. J.

EUGENICS

THE SCIENCE OF HUMAN IMPROVEMENT BY BETTER BREEDING

I. FIT AND UNFIT MATINGS
II. A PLAN FOR FURTHER WORK

I. FIT AND UNFIT MATINGS [1]

THERE comes a time in the life of most thoughtful, cultured people when they realize that they are drifting toward marriage and when they stop to consider if the proposed union will lead to healthful, mentally well-endowed offspring. But however much such a person may take advice of books or friends he will find such a lack of definite knowledge that, shutting his eyes to possible disaster, he decides to take the chances. Were our knowledge of heredity more precisely formulated there is little doubt that

[1] Read, by invitation, before the American Academy of Medicine, at Yale University, Nov. 12, 1909.

3

many certainly unfit matings would be prevented and other fit matings, that are avoided
through false scruples, would be happily con-
tracted. I propose briefly to consider what is
the present state of our knowledge of the inheritance of various characteristics.

The limitations in the scope of this booklet
must be made clear at the outset. As a biologist,
not a physician, I shall not consider many acquired conditions which render unfit for marriage. Governments spend scores of thousands
of dollars and establish rigid inspections to
prevent the spread of the coitus disease of the
horse but the Spirochete parasite that causes the
corresponding disease in man and entails endless
misery on hundreds of thousands of innocent
children may be disseminated by anybody, and
is being disseminated by scores of thousands of
persons in this country, unchecked, under the
protection of the "personal liberty" flag. Alas!
that so little thought is had to the loss of liberty
of the infected children. Marriage of persons
with venereal disease is not only unfit; it is a
hideous and dastardly crime; and its frequency

4

would justify a medical test of all males before marriage, innocent as well as guilty. Fortunately there exists for syphilis at any rate a test so simple that there can be no more objection on any sentimental ground to it than to vaccination.

Nor do I propose to consider in any detail the effects of drugs on germ-plasm. The matter awaits further investigation. Meanwhile experience indicates that the marriage of alcoholists certainly and probably of users of any drug to extremes is associated with defective development of offspring and is, in so far, unfit. Also the class of cases in which, as in tuberculosis, a weakened person is quickly finished by the drain of reproductive processes bears on marriage fitness but does not belong to my topic in the narrow limits I assign it. For my topic deals rather with the result of union of two uninfected germ plasms with their *inherent* peculiarities.

Under these limitations, then, I may say that recent developments in the study of heredity, commonly associated with the name of Mendel, enable us to formulate more precisely than hitherto the working of heredity. Three fundamental principles are to be kept clearly in mind. The principle of independent unit characters, the principle of the determiner in the germ-plasm, and the principle of segregation of determiners.

The principle of independent unit characters states that the qualities or characteristics of organisms are, or may be analyzed into, distinct units that are inherited independently. It follows that the characters of a parent or a particular relative are not inherited as a whole but each individual is a mosaic of characters that appear in a variety of relatives.

The principle of the determiner in the germ-plasm states that each unit character is represented in the germ by a molecule or associated groups of molecules called a *determiner*. These determiners are transmitted in the germ-plasm and are the only things that are truly inherited. It is a corollary of the theory of inheritance from the determiner that we do not inherit from our parents, grandparents or collaterals, but related individuals have some common characteristics because developed out of the same germ-plasm with the same determiners. A child resembles his father because he and his father are developed from the same stuff. Both are chips from the *same* old block. ⟨In relation to determiners some characteristics are positive, depending directly upon them; while others are negative and depend upon the absence of a determiner.⟩ Thus a brown eye depends on an enzyme that produces the sepia-colored pigment, while a blue eye depends upon the absence of such an enzyme. ⟨It is not always easy to anticipate whether a given characteristic is positive or negative.⟩ For instance, long hair as in angora cats, sheep or guinea pigs is apparently not due to a factor added to short hair but rather to the absence of a determiner that stops growth in short haired animals.

The principle of segregation of determiners in the germ-plasm states that characteristics do not blend. That if one parent has a characteristic while the other lacks it, then the offspring get a determiner from one side only instead of from both sides and when the germ-cells are formed in such offspring half of them have the determiner and half of them lack it. There is thus a segregation of presence and of absence of the determiner in the germ-cells of the mixed off-spring. The characteristic in the offspring that is due to a single (instead of the normal double) determiner is called a simplex characteristic. Such a characteristic is frequently distinguishable from one that is due to the double determiner by its imperfect development. Thus the offspring of a pure black-eyed and a blue-eyed parent will have brown eyes.

It is a corollary of the foregoing that if the individual with a simplex character be mated to one lacking the character half of the offspring will lack the determiner and half will be simplex, again, in respect to the character. If in both parents the character be simplex, then two like determiners will meet in one-fourth of the unions of egg and sperm, the two will both be absent in one-fourth of the unions, and one only will occur in half of the unions,—such will be simplex again. If one parent have the characteristic simplex and the other duplex, then half of the offspring will have it simplex and half duplex.

Starting with the principles just enunciated we reach at once the most important generalization of the modern science of heredity:—*When a determiner of a characteristic is absent from the germ-plasm of both parents (as proved by its absence from their bodies) it will be absent in all of their offspring.* In order to predict the result of a particular mating it is necessary first to know what similar unit characteristics both the parents lack, what they both possess and in which characters they differ, and, secondly, to know for each characteristic whether it is due to the presence of a determiner or to its absence. This can, in part, be determined experimentally or inferred from pedigrees. Nevertheless our knowledge of determiners progresses slowly; for here, as in other branches of science, nature's secrets have to be forced from her.

To illustrate the precision with which the characteristics of offspring may be predicted in the best-studied cases, I may refer to eye color. Blue eyes are due to the absence of brown pigment. If there is a determiner for brown iris pigment in the germ-plasm it will produce such

9

pigment in the body that arises from that germ-plasm. The absence of iris pigment is proof of the absence of the pigment determiner from the germ-plasm. If both parents lack brown pigment, their offspring, being devoid of the determiner for brown pigment, will all lack brown pigment. As a matter of experience two parents both with pure blue eyes will have only blue-eyed offspring. Similarly, if the hair of the parents be flaxen, that is evidence of the absence of a hair-pigment determiner in their germ-plasm. In the united germ-plasms of two flaxen-haired parents there is no determiner for hair pigment and all children will have flaxen hair. This agrees, again, with experience. For the same reason parents both lacking curliness or waviness of hair will typically have only straight-haired children.

Hair and eye color are characteristics which serve well to illustrate the precision of the modern science of heredity, but they are ordinarily considered to be immaterial to well-being. But if it is true, as Major C. E. Woodruff maintains, that pigmentation protects individuals from the injurious effects of the tropical sun's rays then one may say that the marriage of two blue-eyed persons in the tropics would be an unfit marriage. On the other hand, the marriage of a blond with a brunet would be fit, for the darker consort would bring into the combination the determiner for pigment and ensure a dark progeny. In the tropics, then, the marriage of light with dark or of two dark persons is, by hypothesis, a fit mating while that of two blonds is unfit.

We may now extend the study of the method of inheritance to cases of abnormalities and diseases, and we shall see that just as it is hard to draw the line between these two sorts of characteristics so they show no difference in their general method of inheritance.

A typical example of an abnormality is that of brachydactyly or short-fingeredness, a condition in which each digit comprises only two phalanges—the fingers are all thumbs. This result seems to be due to an inhibition of the normal growth process. An abnormal person married even to a normal will beget 100 percent or 50 percent abnormal, according to circumstances, and such a marriage is unfit; but two parents who, though derived from brachydactyl strains, altogether lack the inhibitor of growth will have only normal children, for normality implies entire absence of the determiner that stops the growth of the fingers. Such a union is entirely fit.

The rule that the abnormal condition is *induced* by something, so that normal parents never produce abnormal offspring, holds for many abnormal conditions such as presenile cataract; the congenital thickening of the skin known as keratosis; xanthoma, in which the skin acquires yellow patches; hypotrichosis congenita familiaris, or early absence of hair, and other abnormal conditions of skin and hair whose inheritance has been analyzed by Gossage. Probably the same is true of diabetes insipidus and stationary night blindness, according to Nettleship. In all these cases the intermarriage of normal descendants, even of abnormal ancestry, is entirely fit; but abnormals will reproduce their peculiar condition.

In another class of cases the abnormal or dis-
eased condition is due to the *absence* of a characteristic or quality. Thus albinism is due to
the absence of pigment and two albino parents
have only albino children. Normal offspring of
an albino and a pigmented parent may transmit
the albinic condition; and the marriage of a pigmented male of an albinic strain with the pigmented female of another albinic strain or with
a pigmented cousin is unfit. In the case of the
degenerative disease of the retina known as retinitis pigmentosa normals may carry the disease
so that two normal cousins from retinitis stock
may have offspring with retinitis. In fact, a
large percentage of all cases of retinitis come
from consanguineous marriages. Surely such
marriages are highly unfit.

Deafmutism is due to a great variety of
causes; any one of a variety of defects may produce it. But in different individuals of the same
family the chance is large that it is due to the
same defect. This defect is frequently recessive,
hidden in the normal children. Two such normal children who are cousins and from deaf-

mute stock tend to have about one-quarter of their offspring deaf-mutes. (The proportion of congenitally deaf offspring is thrice as great among cousin-marriages as among others.) The conclusion of Fay, based on extensive statistics, deserves to be widely known: " Under all circumstances it is exceedingly dangerous for a deaf person to marry a blood relative, no matter whether the relative is deaf or hearing, nor whether the deafness of either or both or neither of the partners is congenital, nor whether either or both or neither have other deaf relatives besides the other partner.'' Such a marriage has proved to be unfit.

Passing next to the group of mental diseases we find several forms which seem to be due to the absence of some simple unit, so that when both parents exhibit the abnormality all of the children do likewise. As a first case may be taken imbecility

That imbecility is due to the absence of some definite simple factor is indicated by the simplicity of its method of inheritance. (Two imbecile parents, whether related or not, have only

A

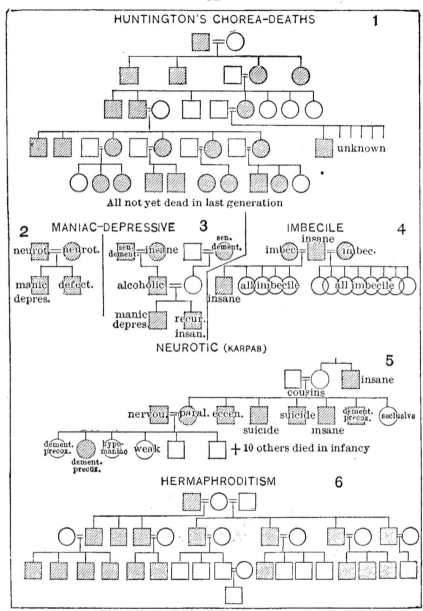

Squares, males ; circles, females ; shaded, affected.

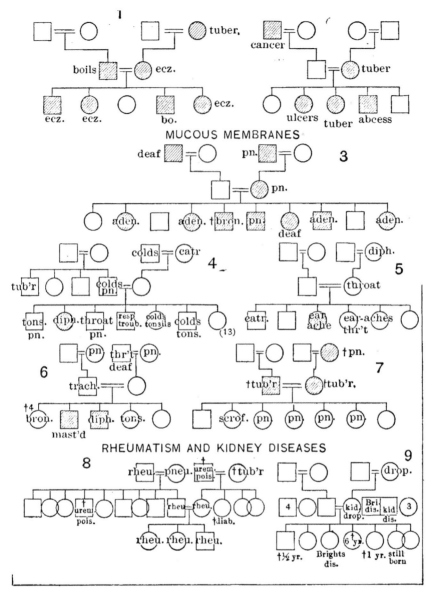

Squares, males; circles, females; shaded or lettered, liable to given disease. *aden.*, adenoids; *bo.*, boils; *Bri. dis.*, Bright's disease; *catr.*, catarrh; *diab.*, diabetes; *drop*, dropsy; *mast'd*, mastoiditis; *pn.* or *pneu.*, pneumonia; *rheu.*, rheumatism; *trach.*, trachitis; *tub'r*, tuberculosis. (13), age.

C

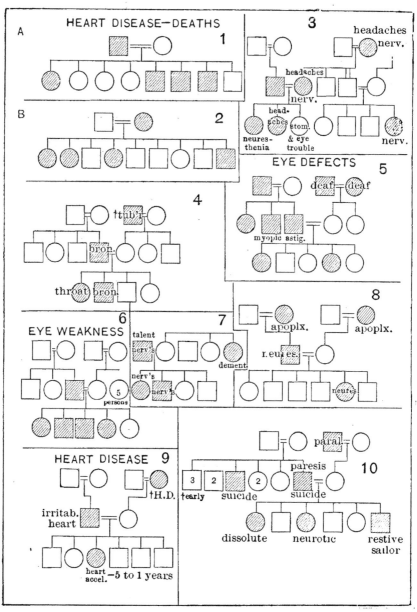

HEART DISEASE—DEATHS

A — 1

B — 2

3 — headaches nerv.
headaches nerv.
head-aches
neures-thenia / stom. & eye trouble / nerv.

EYE DEFECTS — 5
deaf — deaf
myopic astig.

4 — †tub'? bron. throat bron.

EYE WEAKNESS — 6
talent nerv's
nerv's nerv's dement.
5 persons

7 — apoplx. r.eures.
8 — apoplx. neures.

HEART DISEASE 9
†H.D.
irritab. heart
heart accel. — 5 to 1 years

10 — paral. paresis
paresis suicide
3 | 2 | 2 †early suicide
dissolute neurotic restive sailor

Squares, males ; circles, females ; shaded, affected. Figures in circles indicate number of normals. *Astig.*, astigmatism ; *bron.*, bronchitis ; H. D., heart disease ; *nerv.*, nervous.

D

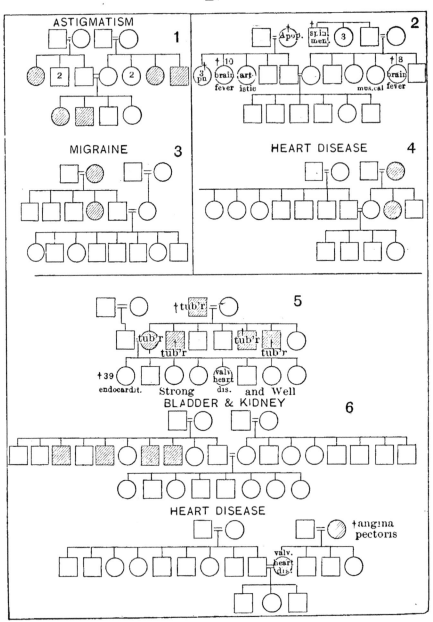

Squares, males; circles, females; shaded, affected. Figures *in* circles
indicate 2 or 3 normals; † 3, † 8, † 39, age at death. *Apop.,*
 nlosis

(imbecile offspring.) Barr gives us such data as the following from his experience. A feeble-minded man of 38 has a delicate wife who in 20 years has borne him 19 defective children. A feeble-minded epileptic mother and an irresponsible father have 7 idiotic and imbecile children. The L family numbers 7 persons, both parents and all 5 children imbecile. Among the " Family Records " I have been collecting there occurs the R family where A (insane) marries in succession two mentally weak wives and has 13 children, all mentally weak. (Plate A 5.) In a case described by Bennett, a defective father and imbecile mother have 7 children all more or less mentally and morally defective. There is, so far as (I am aware, no case on record where two imbecile parents have produced a normal child. So definite and certain is the result of the marriage of two imbeciles, and so disastrous is reproduction by an imbecile under any conditions that it is a disgrace of the first magnitude that thousands of children are annually born in this country of imbecile parents to replace and probably more than replace the deaths in the army

15

of about 150,000 mental defectives which this country supports. The country owes it to itself as a matter of self-preservation that every imbecile of reproductive age should be held in such restraint that reproduction is out of the question. If this proves to be impracticable then sterilization is necessary—where the life of the state is threatened extreme measures may and must be taken.

Maniac-depressive insanity seems likewise due to a defect, in any case it is especially apt to occur in families in which both parental strains show the disease. I give a few cases. (A 2, 3.)

While, on account of the complexity of nervous diseases, all of the children even of two neurotic parents are not always neurotic, the chances of this result are much increased when the parents are related. This is illustrated by the family described by Karpas. Here all children are nervously defective. (A 4.) 5.

The case of partial hermaphroditism is peculiar because it affects usually only the male sex. The inhibitor of complete sex differentiation seems to be dominant in the male—the embryologically more advanced sex—though it may fail to activate in and is indeed irrelevant to the female sex. (A 6.) Since the abnormality is necessarily revealed only by the male sex the condition of the female is no test of her germ-plasm in respect to this characteristic. As a matter of fact the normal mother may easily represent the defective strain. A normal male belonging to the defective strain is usually with-out trace of the inhibitor, yet a few cases are known of an apparently normal person with an inactive '' inhibitor '' having, by a normal con-sort, some abnormal sons. But, in general, the marriage of females belonging to hermaphrodite (hypospadie) strains is unfit, while normal males of such strains may marry females from normal strains.

The case of Huntington's chorea is a striking one of inheritance of disease. This is a form of chorea that leads to dementia and death.) A. S. Hamilton has worked up the pedigree of many cases. (A 1.) The mating of two parents with chorea is obviously highly unfit and should not be permitted.

Let us now consider the hereditary behavior of some of the commoner diseases, including those which, while not fatal or apt to incapacitate a person, nevertheless interfere much with his happiness. Knowledge in this field is less precise, although the general teaching is not less clear. As a source of information I rely chiefly on the records of health and other characteristics furnished for over 200 families by members of the families concerned. These are largely representatives of professional circles, but include also farmers and people in commercial life.

In the pedigrees that follow nothing is more evident than that usually specific diseases are not inherited but only a condition of liability or non-resistance to a particular class of disease. Often an entire organ-complex is thus non-resistant.

A good example of inheritance of general weakness in an organ is sometimes found in the case of the mucous membranes.

In the N family the principal diseases to which there is liability are located in the mucous membranes of nose, throat, ear and bronchia. (B 3.)

In the D family the center of susceptibility is more specific, being nearly confined to the nose and throat. (B 4.)

In the E family the center of weakness is the ear. (B 5.)

In the N family the trouble seems to spread from the throat. (B 6.)

In the M family the susceptibility is more nearly confined to the lungs. (B 7.)

In another family the *skin* seems to be the weak organ, boils and eczema are common. (B 1.) In still another tubercles and abscesses seem to be associated. (B 2.)

In other families the kidneys will be the seat of incidence. In one it will take the form of Bright's Disease and dropsy (B 9): in another uremic poisoning and "rheumatism." (B 8.)

Heart disease is a very general term; there is no doubt about its inheritance although the precise nature of the weakness is varied. (C 1, 2.)

Nervous disease re-appears as paralysis, neurasthenia, nervousness, headaches and stomach trouble, and migraine appears in successive generations. (C 3, 10; D 2.)

Looking over 'these pedigrees one is impressed by the fact, first, that the incidence of diseases is . not haphazard nor, in any large family, do the various causes of death occur in the proportions given in the census tables for the population as a whole. Tuberculosis of the lungs is the cause of more than 10 percent of the deaths in the United States but it would not be difficult to pick out of my collection ten families comprising about 100 deceased persons among whom, instead of the expected 10 not one died of consumption. Similarly there are many families in which no nervous disease has occurred in three generations; others without kidney diseases and so on. On the other hand, in other families 40 to 50 percent or even 80 percent are attacked by lung and throat troubles or nervous defects. These differences cannot be attributed chiefly to environment, because they occur in families of which the members are widely dispersed and have varied occupations. They indicate fundamental differences in the protoplasm.

But, it may properly be urged, how about environment? Are not many of these examples of

20

occurrence of disease in one family due to infection or to similar untoward conditions? I do not doubt that they *all* are. The controversy between heredity vs. environment has no good basis and it is fallacious to emphasize the distinction. As well might one ask whether poor seed or poor climate is the more important in determining poor crops; both are important. Nevertheless, emphasis must be laid on the fact that while poor climate brings heavy losses, there are strains that you can hardly kill by frost, nor by drought, nor by poor soil, nor by the wilt parasite—there is such a thing as resistance in the blood as well as, on the other hand, *susceptibility* of particular organisms to poor environment or to infection. Unfavorable environment collects its toll first from those who are, by heredity, least resistant.

Granting the fundamental fact of the diversity in resistance or liability to disease of the different protoplasms it remains to be considered how these facts are to be applied in selecting consorts so as to secure healthy children.

In some cases at least definite rules may be

laid down. The fundamental law, is, as already stated: Whenever the same unit defect exists in both parental protoplasms it will appear in all the offspring. The " unit defect " is not, as already pointed out, easily determined, nor is a given gross defect probably identical in the parental strains unless the parents are cousins. Despite these difficulties in its application the rule holds, by and large, as a valuable first approximation.

Of unit defects the weakness of mucous membranes seems to be a good illustration. If both parents are subject to colds, catarrh, bronchitis, asthma or lung-tuberculosis all or nearly all of the children are liable to these diseases. The same is true in some forms of nervous disease and rheumatism. If the disease fails to appear in any child it is probably because the child died too early, or is still very young, or has been able through exceptionally favorable environment to avoid the incidence of the disease or, by strengthening other means of defence, to hold it down or eliminate it, even after attack. The expectation that is usually realized, however, is that all shall show a weakness to the same disease as their parents.

If liability to the disease is found in the protoplasm of both parental strains but is shown in the soma of one only of the parents, then it will probably occur in one-half of the offspring. Examples are found in the families whose pedigrees are given in the diagrams. (C 5; B 3; C 4; C 7; C 10; C 8, C 9.) The total of the last generation in these examples gives 18 subject to the specific disease and 23 non-subject where $20\frac{1}{2}$ and $20\frac{1}{2}$ are expectation. The excess of the non-subject may be explained on the same ground as the exceptions to complete incidence of disease referred to in the preceding paragraph.

If both parents belong to strains having the same unit defect even though they have it not themselves we may expect either that one-quarter or none of the children will have the defect, depending on earlier ancestry. This rule is illustrated by some of my cases. (D 1, D 3.)

If one parent belongs to a strain with a unit defect while the other strain is without the defect then the children will be without the defect.

This is illustrated by many examples. (D 2, D 4, D 5, D 6, D 7.)

To the rule that a strong characteristic from one strain may overcome the defects of a weak characteristic from the other strain there are some apparent exceptions; due chiefly to the fact that the simplex condition is rarely quite as strong as the purely positive condition so that defects are not wholly overcome and to the fact that the supposed pure positive strain may contain a hidden defect and be really only simplex.

Recognizing these limitations in our knowledge, which it is believed further accumulation and study of data will overcome, how far can we go in advising, in the case of the commoner hereditary diseases, what matings are and what are not conducive to healthy offspring?

The foregoing considerations indicate: If (A) the *negative* character is, as in polydactylism and night blindness, the *normal* character, then normals should marry normals and they may even be cousins. B. If the negative character is *abnormal,* as in imbecility and liability to respiratory diseases, then the marriage of two abnormals means probably all children abnormal; the marriage of two normals from defective strains means about one-quarter of the children abnormal, but the marriage of a normal of the defective strain with one of a normal strain will probably lead to strong children. The worst possible marriage in this class of cases is that of cousins from the defective strain, especially if one or both have the defect. In a word, the consanguineous marriage of persons one or both of whom have the same undesirable defect is highly unfit, and the marriage of even unrelated persons who both belong to strains containing the same undesirable defect is unfit. Weakness in any characteristic must be mated with strength in that characteristic; and strength may be mated with weakness.

II. A PLAN FOR FURTHER WORK [1]

IN what I have just said I have tried to be cautious, and have felt, at every step, that generalizations are restricted by lack of sufficient facts. This is the serious need of the time. To fill this need the American Breeders' Association has organized a Committee on Eugenics composed as follows: David Starr Jordan, Chairman; Alexander Graham Bell, Luther Burbank, W. E. Castle, C. R. Henderson, A. Hrdlicka, V. L. Kellogg, Adolf Meyer, J. Arthur Thomson, W. L. Tower, H. J. Webber, C. E. Woodruff, Frederick A. Woods, C. B. Davenport, Secretary. The various duties of this Committee may be summed up in the three words: investigation, education, legislation.

The first, and for some time the main work of the Committee must be *investigation*. We want, above all, to learn as soon as possible how human

[1] This part is substantially from a report of the Committee of Eugenics of the American Breeders' Association, read at the Omaha meeting, Dec. 8, 1909.

26

characteristics are inherited. The results of the new science of heredity give reason for anticipating that many, if not most, characteristics are of an alternative sort either not re-appearing in the offspring or re-appearing in predictable proportions depending upon the distribution of these characteristics in the ancestry. We have already seen that a score or more of characteristics, largely specific diseases, are inherited in such alternative fashion, and about their behavior in progeny definite information has been given. We must ascertain the facts about other characteristics.

The data must first be collected; then analyzed. This work is so vast that it must be divided between many people,—specialists able to weigh and analyze scientifically the results. Consequently it has been found desirable to appoint sub-committees to collect and study the data. A sub-committee on Feeble-mindedness has been organized under the chairmanship of Dr. A. F. Rogers, Superintendent of the Minnesota School for Feeble Minded and Colony for Epileptics, and with Dr. H. H. Goddard, Director of the Department of Psychological Research at the New Jersey Training School for Feeble Minded Girls and Boys, as Secretary. At the present moment this committee is collecting answers to the question: "Do two imbecile parents ever beget normal children?" This committee has most important interests, since the number of feeble-minded in the United States alone is probably not less than 150,000 of which 15,000 are in institutions.[1]

A sub-committee on Insanity is being organized under the chairmanship of Dr. Adolf Meyer, for some time Director of the Pathological Institute of the New York State Commission in Lunacy and recently appointed head of the Phipps Psychiatrical Institute at the Johns Hopkins University The secretary of this Sub-Committee is Dr. E. E. Southard, Pathologist to the State Board of Insanity, Massachusetts. This sub-committee has important work to do, for there are over 150,000 insane in the Institutions of the United States alone.

[1] Bureau of the Census. Special Reports : Insane and Feeble-minded in Hospitals and Institutions. 1904, p. 205.

Other sub-committees are contemplated to study the protoplasmic basis of eye defects, deafness, predisposition toward lung and throat trouble, toward diseases of the excretory and circulatory organs; toward cancer, skin diseases, crippled appendages and so on. Still other sub-committees should deal with criminality and pauperism, with the effects of consanguineous marriages and of such mongrelization as is proceeding on a vast scale in this country. Perhaps other sub-committees, recruited from those who make physical examinations, will study inheritance of muscular strength, of sound wind and endurance. Possibly registrars of colleges will serve on sub-committees for the study of inheritance of various intellectual traits. Other sub-committees will be added as needed.

A second class of investigation may better be undertaken by the central committee. It is the obtaining of records of the inheritance of char acteristics of health, ability and temperament from typical American families. In the attempt to secure such records 5000 blanks have been distributed and about 300 Family Records received back. These are being studied to determine the laws of incidence of disease and the inheritance of various other characteristics. This sort of work might be taken up by genealogists who wish to incorporate more biological data in their family histories. The limitations to this work are set only by lack of means for carrying on correspondence. It seems possible that data of this sort might be collected by the national Bureau of the Census for limited registration areas.

While the acquisition of new data is desirable, much can be done by studying the extant records of institutions. The amount of such data is enormous. They lie hidden in records of our numerous charity organizations, our 42 institutions for the feeble-minded, our 115 schools and homes for the deaf and blind, our 350 hospitals for the insane, our 1200 refuge homes, our 1300 prisons, our 1500 hospitals and our 2500 almshouses. Our great insurance companies and our college gymnasiums have tens of thousands of records of the characters of human blood lines. These records should be studied, their hereditary data sifted out and properly recorded on cards and the cards sent to a central bureau for study in order that data should be placed in their proper relations in the great strains of human protoplasm that are coursing through the country. Thus could be learned not only the method of heredity of human characteristics but we shall identify those lines which supply our families of great men: our Adamses, our Abbotts, our Beechers, our Blairs, and so on through the alphabet. We shall also learn whence come our 300,000 insane and feeble-minded, our 160,000 blind or deaf, the 2,000,000 that are annually cared for by our hospitals and Homes, our 80,-000 prisoners and the thousands of criminals that are not in prison, and our 100,000 paupers in almshouses and out.

This three or four percent of our population is a fearful drag on our civilization. Shall we as an intelligent people, proud of our control of nature in other respects, do nothing but vote more taxes or be satisfied with the great gifts and bequests that philanthropists have made for the support of the delinquent, defective and dependent classes? Shall we not rather take the steps that scientific study dictates as necessary to dry up the springs that feed the torrent of defective and degenerate protoplasm?

Greater tasks than those contemplated in the broadest scheme of the Eugenics committee have been carried out in this country. If only one-half of one percent of the 30 million dollars annually spent on hospitals, 20 millions on insane asylums, 20 millions for almshouses, 13 millions on prisons, and 5 millions on the feeble-minded, deaf and blind were spent on the study of the bad germ-plasm that makes necessary the annual expenditure of nearly 100 millions in the care of its produce we might hope to learn just how it is being reproduced and the best way to diminish its further spread. A *new* plague that rendered four percent of our population, chiefly at the most productive age, not only incompetent but a burden costing 100 million dollars yearly to support would instantly attract universal attention, and millions would be forthcoming for its study as they have been for the study of cancer. But we have become so used to crime, disease and

32

degeneracy that we take them as necessary evils. That they were, in the world's ignorance, is granted. That they must remain so, is denied.

The second great duty of the Committee on Eugenics, education, is not less important than investigation. For the ascertained laws would be more than scientifically interesting; they would be guides to action on the part of the reading, thinking public. As precise knowledge is acquired it must be set forth in popular magazine articles, in public lectures, in addresses to workers in social fields: in circular letters to physicians, teachers, the clergy and legislators. The nature and the dangers of unfit matings, the way to secure sound progeny, must ever be set forth.

And, finally, when public spirit is aroused, its will must be crystallized in appropriate legislation. Since the weak and the criminal will not be guided in their matings by patriotism or family pride, more powerful influence or restraints must be exerted as the case requires. And as for the idiots, low imbeciles, incurable and dangerous criminals they may under appropriate

restrictions be prevented from procreation—
either by segregation during the reproductive
period or even by sterilization. Society must
protect itself; as it claims the right to deprive
the murderer of his life so also it may annihilate
the hideous serpent of hopelessly vicious protoplasm. Here is where appropriate legislation
will aid in eugenics and in creating a healthier,
saner society in the future.

We come now to the practical question, how
can the necessary studies be made? It is believed that the Committee of Eugenics may well
be entrusted with organizing the work along the
lines that have been successfully begun or it
would co-operate with any body that seemed better able to organize the work. But it can do
nothing without funds. The committee does not
solicit funds—but it stands ready to do the nation's business by making clear the nation's need
to legislators and to philanthropists. One cannot fail to wonder that, where tens of millions
have been given to bolster up the weak and alle-
viate the suffering of the sick, no important
means have been provided to enable us to learn

how the stream of weak and susceptible proto-plasm may be checked. Vastly more effective than ten million dollars to " charity " would be ten millions to Eugenics. He who, by such a gift, should redeem mankind from vice, imbe-cility and suffering would be the world's wisest philanthropist.

Playing God + the effects

Trying to rid all abnormalities from human genes by firstly human experimentation then assumingly alimination.

lots of self ~~prof~~ proclaimed geniousness - Significance - narcissmism -

lack of emotion or humanness Shows Sign of Some form of mental abnormality within itself

like Schizophrenia

Putting himself in a higer rank - as well as other Eugenisists wald have.

Wald encavrage People heaving children seeing them as

not perfect so scrap + try
again how this behaviour
would remove emotion
and positive meaning
leading us to an
apathetic / lack of disgust
for behaviours type of
world. Therefore by trying
to rid ill health, Mental
issues and criminality would
be a paradox for the type
of world that would be
Created.

Eugenics in this way leads
to people of insanity hiding their
insanity in the disguise of
the greater good. →

Saying this how do we take the idea of eugenics in and work on more individuals taking pro-creation more seriously in the way of health, mental health, life style, their socialisedness etc etc to create healthier homes, families & societies – it being used metaphorically in our minds ruling out if we are Printed in Great Britain by Amazon healthy enough to have children or who we are attracted to is – how we can work